W9-CGW-765

E=MC2
LEARN
Dream BIG,
Little Scientists
RECORD

To Mom and Dad, for nurturing my curiosity.—M. S.

To all future scientists: be curious, always!—A. P.

Published by Charlesbridge
9 Galen Street
Watertown, MA 02472
(617) 926-0329
www.charlesbridge.com

Library of Congress Cataloging-in-Publication Data
Names: Schaub, Michelle, author. Potter, Alice, illustrator.
Title: Dream big, little scientists: a bedtime book / Michelle Schaub; illustrated by Alice Potter.
Description: Watertown, MA : Charlesbridge, [2020]
Identifiers: LCCN 2018052243 (print) LCCN 2018054965 (ebook) ISBN 9781632897862 (ebook) ISBN 9781632897879 (ebook pdf) ISBN 9781580899345 (reinforced for library use)
Subjects: LCSH: Science—Miscellanea—Juvenile literature. Children's poetry. Lullabies.
Classification: LCC Q163 (ebook) LCC Q163 .S33 2020 (print) DDC 500—dc23 LC record available at https://lccn.loc.gov/2018052243

Printed in China
(hc) 10 9 8 7 6 5 4 3 2

Illustrations done in Adobe Illustrator
Display type and text set in Eat Well Chubby by Chank Co
Hand-lettering by Alice Potter
Color separations by Colourscan Print Co Pte Ltd, Singapore
Printed by 1010 Printing International Limited in Huizhou, Guangdong, China
Production supervision by Brian G. Walker
Designed by Joyce White & Jacqueline Noelle Cote

Michelle Schaub ◦ *Illustrated by* Alice Potter

Dream BIG, Little Scientists

Charlesbridge

Dream BIG,
little scientists,

YAWN!
SIGH!
READY FOR BED
SO TIRED

SO SLEEPY
SIGH!
SIGH!
YAWN!

and close your sleepy eyes. . . .

SPACE
ASTRONOMY

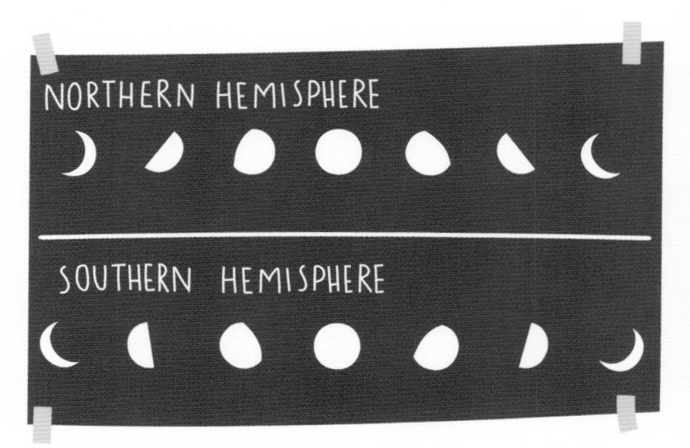
NORTHERN HEMISPHERE
SOUTHERN HEMISPHERE

you are a STAR

The sun has tucked itself in bed;

CARL SAGAN

STARS & PLANETS
SKY AT
ASTRONOMY MAGAZINE

STARGAZER 3000

the moon is on the rise.

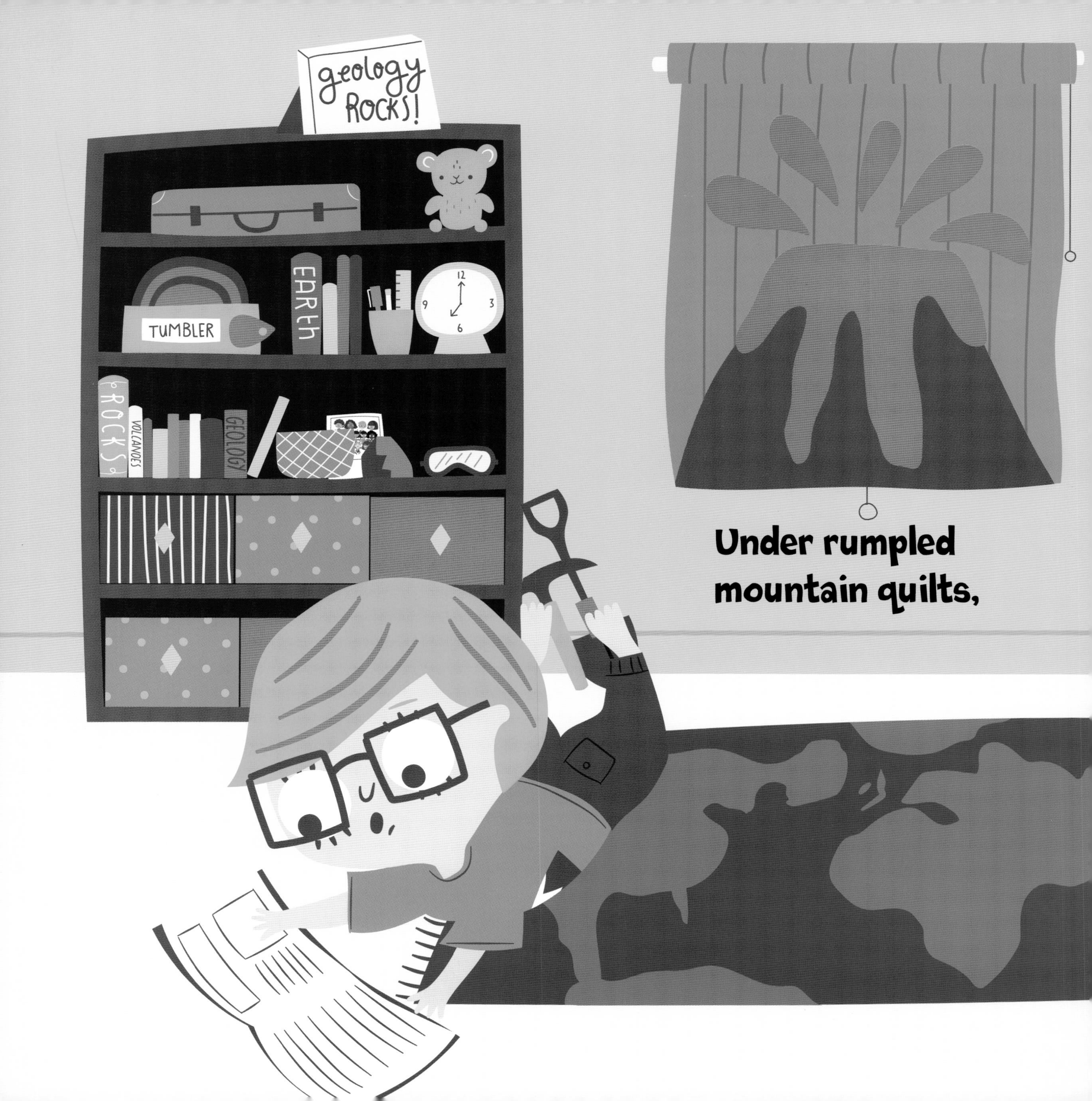
geology ROCKS!
TUMBLER
EARth
ROCKS
VOLCANOES
GEOLOGY

Under rumpled
mountain quilts,

ROCKS

JESS PHOENIX

the earth is snuggled tight.

The oceans rock the world to sleep;

JACQUES COUSTEAU
Oceanography
WHALE SONGS VII
BLUE PLANET

sunlight
twilight
midnight
ABYSS
TRENCH

the waves whisper,
"Good night."

CLIMATE CHANGE
IS
HAPPENING
RIGHT
NOW

RAIN
tornado

TODAY IS

Rain and snow and winds that blow

CLIMATE AND METEOROLOGY

all hum a lullaby.

GABRIEL FAHRENHEIT
ANDERS CELSIUS
BAROMETER
60
54
48
PHOEBE
42
ROSE
36
30
24
18
12
6
0

GEORGE WASHINGTON CARVER
THOMAS MEEHAN
A FLOWER
BOTANY

While mossy carpets stretch out wide,

PLANTS AND TREES
PLANTS AND TREES
FLOWERS
SEEDS

tree limbs yawn up high.

EXPLORER
ECOLOGY
VOL 1
GRASSLANDS
SAFARI
DESERTS

EARTH
FORESTS
ANIMALS

Daytime creatures
settle down

DESERT
GRASSLAND
OCEAN
AIN FOREST
TUNDRA
TAIGA

WANGARI MAATHAI

in den or hole or nest.

As motion slows and quiet grows,

ISAAC NEWTON
LAW
MOTI
FLIGHT
UP&
DOW
PHYSICS
100 QUESTIONS
SCIENCE
VITY
DONNA
STRICKLAND
E=MC²

objects come to rest.

ΣF=ma
STEPHEN HAWKING

THE JURASSIC PERIOD
Dear museum directo
ROAR
CURRENT PROJECTS
FUTURE PROJECTS
STUFF

Slumber's been a
part of life

since prehistoric days.

MARY ANNING
FOSSIL KIT
BIG DINO BOOK
PALEONTOLOGY

MIZUKO ITO
ANTHROPOLOGY
HUMANS
HUMAN MIGRATION
CULTURES
STORY OF MAN
BIG BOOK OF TRIBES

And people all around the world

EVOLUTION OF
HUMANS
TAKIE LEBRA

bed down in different ways.

Breathe in deep,
then let it out,

JOSEPH LISTER
DR
MEDICINE
HUMAN BODY
HEALTH
DISEASE
PHYSIOLOGY
MEDICINE
I ♥ MY
BODY

and feel your heartbeat slow.

Like water flows
from melting ice,

MARIE CURIE
ALICE BALL
ELEMENTS
CHEMISTRY KIT

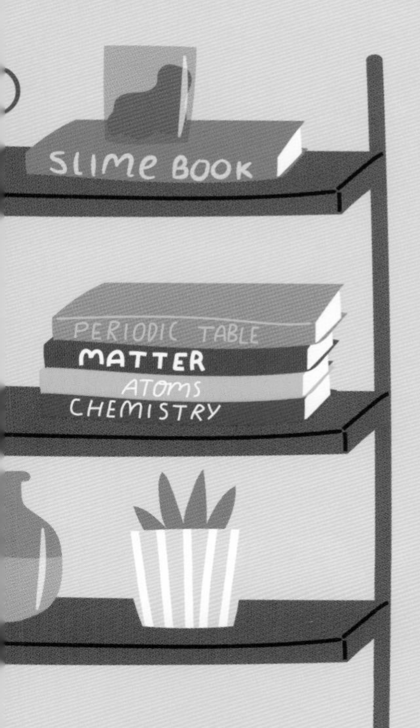
SLIME BOOK
PERIODIC TABLE
MATTER
ATOMS
CHEMISTRY

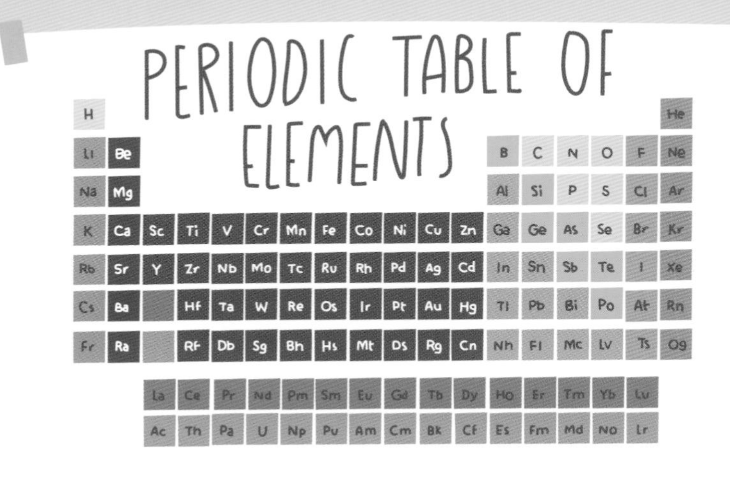
PERIODIC TABLE OF ELEMENTS
H He
Li Be B C N O F Ne
Na Mg Al Si P S Cl Ar
K Ca Sc Ti V Cr Mn Fe Co Ni Cu Zn Ga Ge As Se Br Kr
Rb Sr Y Zr Nb Mo Tc Ru Rh Pd Ag Cd In Sn Sb Te I Xe
Cs Ba Hf Ta W Re Os Ir Pt Au Hg Tl Pb Bi Po At Rn
Fr Ra Rf Db Sg Bh Hs Mt Ds Rg Cn Nh Fl Mc Lv Ts Og
La Ce Pr Nd Pm Sm Eu Gd Tb Dy Ho Er Tm Yb Lu
Ac Th Pa U Np Pu Am Cm Bk Cf Es Fm Md No Lr

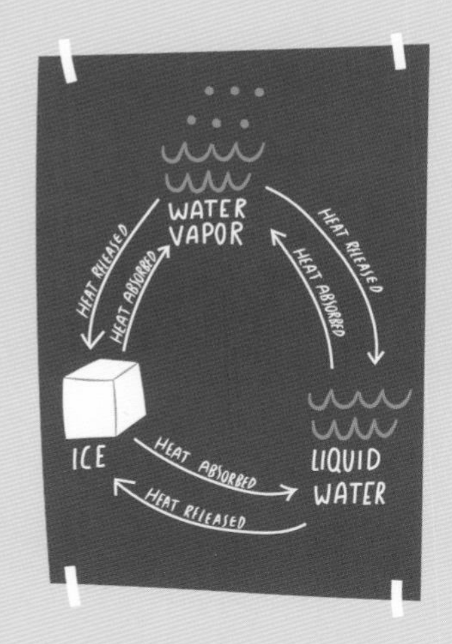
WATER VAPOR
HEAT RELEASED
HEAT ABSORBED
HEAT RELEASED
HEAT ABSORBED
ICE
LIQUID WATER
HEAT ABSORBED
HEAT RELEASED

peace spreads from head to toe.

Dream away,
young scientists,

tomorrow you'll
learn more—

when you awake
and venture out

NAMASTE

to ask, observe, explore.

THINK LIKE A SCIENTIST

Be curious!

Look around, explore, and talk about the world where you live . . . just like a scientist!

To explore the different branches of science even more, visit:

www.sciencekids.co.nz

ASTRONOMY

The sun is a star made of burning gas. It provides light and heat for Earth. The moon looks bright, but it does not make its own light. It reflects light from the sun. Astronomers study space, stars, and planets. Find the moon in the sky. How does it change each night?

GEOLOGY

Earth is covered by layers of rock called the crust. Mountains form when parts of the crust bump into other parts. Geologists study the materials that form our planet. Can you find rocks when you go on walks? Sort them by size, shape, and color.

OCEANOGRAPHY

Oceans cover most of Earth. This is why Earth is called the Blue Planet. Oceanographers study oceans and sea creatures. Visit a pet store or an aquarium to look at fish and other marine animals.

METEOROLOGY

Rain is water that falls from clouds. Snow is frozen water. It falls from clouds, too. Wind is air that moves. Rain, snow, and wind are types of weather. Meteorologists study weather. Watch the wind move clouds across the sky. What other objects can the wind move?

BOTANY

Moss and trees are both plants. Plants make their own food from sunlight. Botanists study plants. Observe leaves, seeds, and flowers outside. How are they alike or different?

ECOLOGY

Every plant and animal has its own place to live called a habitat. Ecologists study how things live and work together. Look closely at a backyard or green space. Make a list of the plants and animals you see.

PHYSICS

People and objects move when forces act on them. A force is any push or pull. Physicists study forces. Drop different-sized ball from the same height at the sam time. Which one do you think will hit the ground first?

PALEONTOLOGY

Some creatures, such as dinosaurs, lived on Earth long ago. They left clues such as shells, bones, and tracks preserved in rock. They are called fossils. Paleontologists study ancient plants and animals. Look for tracks in snow or mud. What animal or plant left them? Visit a museum to see fossils.

ANTHROPOLOGY

People eat different foods, speak different languages, and even sleep differently. Anthropologists study people around the world. Try food from different countries. Look up the countries on a map. What can you learn about the people who live there?

PHYSIOLOGY

Your heart has a special job. It beats to move blood through your body. All parts of the body need blood to work. Physiologists study the body and how it works. Can you feel your heart beating?

CHEMISTRY

Everything is made up of tiny pieces that you cannot see calle atoms. Chemists study how atom behave and change. Watch an ice cube melt in a glass. Put the glas in the freezer—what will happen?

CHEMISTRY KIT

VISIT WWW.MICHELLESCHAUB.COM/SCIENTISTS TO LEARN ABOUT THE SCIENTISTS ON THE POSTERS IN EACH KID'S ROO